Environmental Division 2022

AA009279

Held at the 2022 AIChE Spring Meeting and 18th Global Congress on Process Safety

San Antonio, Texas, USA
10 – 14 April 2022

ISBN: 978-1-7138-6827-9

Printed from e-media with permission by:

Curran Associates, Inc.
57 Morehouse Lane
Red Hook, NY 12571

AIChE
120 Wall Street, FL 23
New York, NY 10005-4020

Phone: (800) 242-4363
Fax: (203) 775-5177

www.aiche.org

Additional copies of this publication are available from:

Curran Associates, Inc.
57 Morehouse Lane
Red Hook, NY 12571 USA
Phone: 845-758-0400
Fax: 845-758-2633
Email: curran@proceedings.com
Web: www.proceedings.com

TABLE OF CONTENTS

Modeling of Carbon Capture Process

Maria Thuy Do
Swalm School of Chemical Engineering,
Mississippi State University
mtd272@msstate.edu

Amin Amirlatifi
Swalm School of Chemical Engineering,
Mississippi State University
amin@che.msstate.edu

Robert Cormia
Physical Sciences, Mathematics and Engineering Division
Foothill College Affiliation
cormiarobert@fhda.edu

Keywords: Carbon Capture, Direct Air Capture, Kinetics, Activated Carbon, Zeolite

Abstract

An adsorbent's desorption kinetics and adsorption capacity are key parameters in planning, modeling, and designing direct air capture (DAC) systems. Determining these parameters reveals the thermodynamics requirements and the properties that favor the CO_2 adsorption process. The present work offers a review of research on the desorption kinetics and adsorption capacity of activated carbon and zeolite for direct air CO_2 capturing. Activated carbon and zeolite samples were studied by thermogravimetric analysis (TGA) with experimental cycles consisting of equilibration, adsorption, and desorption. The kinetic parameters such as the desorption activation energy and pre-exponential factor were obtained by fitting the thermal analysis data. Variable amounts of CO_2 that are representative of DAC processes were used to study the adsorbent's adsorption capacity. The results from the studies suggest that efficient CO_2 adsorption processes are attained when employing higher partial pressure of CO_2.

1 Introduction

Negative Emission Technologies (NETs) are methods used to achieve climate goals by removing and sequestering carbon dioxide (CO_2) from the air. Amongst various NETs, Direct Air Capture (DAC) has captured immense attention for its effectiveness in capturing atmospheric CO_2 and the potential for large-scale implementation [1]. Some mainstream DAC technologies use an array of fans to pass CO_2 across sorbent filters that are amine-based on solid porous granules [2]. In contrast, other commercial DAC players use liquid-solvent-based potassium hydroxide (KOH) as a sorbent to react with calcium hydroxide ($Ca(OH)_2$) in the form of solid pellets to capture atmospheric CO_2 [3].

The DAC technologies face several challenges, including the high energy requirements for the regeneration process, which act as the energy barrier to capture atmospheric CO_2. It is of utmost importance to understand the contribution of the energy barrier and to use the surface characteristics to predict the adsorbent's sorption behavior. While the intensive energy requirements to regenerate the sorbent used for the DAC technologies are inevitable, the attainable kinetic parameters can be used to select viable sorbents and propose a design that effectively reclaims the heat dissipation during the regeneration process.

In this work, adsorbents for DAC such as activated carbon and zeolite are assessed. The primary objective is to understand the adsorbents' performance in terms of desorption kinetics and adsorption capacity. Section 2 discusses the research conducted on the adsorbents. Section 3 discusses the findings and the review on the adsorbents' performance followed by the conclusion in section 4.

2 Literature Review

Surface area and porosity are key attributes to gas adsorption on solid materials. Materials with high surface area and pore volume provide a greater capacity to contain a larger volume of gas to penetrate the material through the available free site on its surface. Adsorption is a consequence of both dispersion and short-range repulsive forces [3]. Porous materials such as activated carbon and zeolites are easily attainable while chemically and physically robust for the carbon capture process [4]. Since the pressure induces surface interaction and adsorbents are temperature dependent, it can be deduced that CO_2 capture capacities are supported by low temperature and high pressure [4]. Activated carbons are porous adsorbents with high content of micro- and mesopore structures. Modifying activated carbon to incorporate heteroatoms such as oxygen, nitrogen, and sulfurs increases the capabilities of gas adsorption by inducing pole-to-pole interaction and the basic site on its surface [4,5]. Zeolite consisting of aluminosilicates with high orders of porosity have high selectivity towards CO_2 by allowing the pass-through of H_2, O_2, and N_2. Several researchers have found zeolites to outperform activated carbon in terms of their DAC performance in low-pressure settings, as well as the superiority of the material's performance for its capability to undergo numerous cycles [6].

2.1 Desorption Activation Energy

Transfer processes of adsorbates onto adsorbents are either caused by physisorption or chemisorption. Physisorption is an adsorption process that occurs due to the van der Waals forces [7]. These reversible processes undergo fast kinetics, as there is no energy barrier to overcome; thus, the activation energy requirements are non-existent [7,8]. The potential energy curve for the physisorption process consists of a shallow minimum, and at a relatively large distance away from the adsorbent's surface, the process is defined as non-activated, resulting in zero potential energy [8]. On the other hand, chemisorption processes involve transferring electrons between the absorbate and absorbents. In contrast to physisorption, chemisorption has to overcome the energy barrier for the reaction to occur, and the adsorbent's surface is more localized than physisorption [8].

Activated carbon and amorphous zeolites have received immense attention as porous carbon adsorbents for DAC technologies. In the study led by Zhao et al., the primary focus was to study temperature swing adsorption (TSA) on activated carbon and zeolite 13X for post-combustion of carbon dioxide capture. A model was generated to study a 4-step TSA cycle for the adsorbents, namely pressurization, adsorption, heating, and cooling, as illustrated in Figure 1 [9].

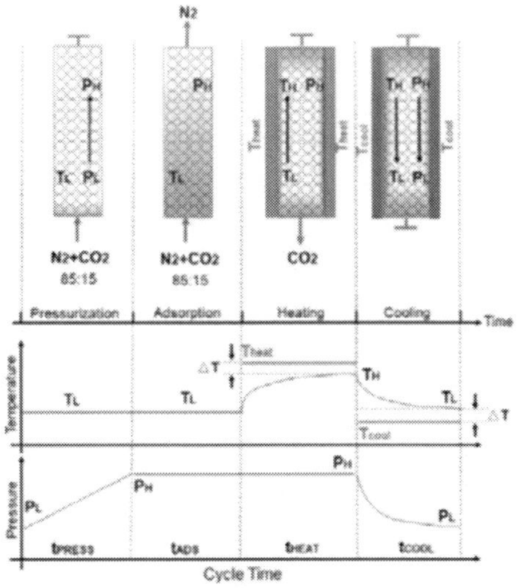

Figure 1. Process Schematic for a Four-Step TSA Cycle [9]

The bed was pressurized to 1 bar while introducing 15% CO_2 and 85% N_2. The adsorption process occurred at a temperature of 303-323°K. Heating was then done at a temperature range of 373-393°K, and the desorbed materials exited from the bottom of the bed. The cooling step proceeded by cooling the adsorbent to 303°K. The isotherm models for the adsorbents are summarized in Table 1.

Table 1. Equilibrium isotherms for the adsorbents [9]

Adsorbents	Isotherm Type	Isotherm Model
Activated Carbon	Toth	$n_i = \frac{q_m K_{eq,i} P_i}{[1 + (K_{eq} P_i)^n]^{\frac{1}{n}}} K_{eq,i} = K_0 \exp(\frac{-\Delta H}{RT})$
Zeolite 13x	Extended Sips	$n_i = \frac{n_{ref,i}(b_i P_i)^{C_i}}{1+\sum 2_{i=1}(b_i P_i)^{C_i}} = n_0 \exp\left(X_i\left(\frac{T}{T_{ref}} - 1\right)\right) b_i = b_{0,i} \exp\left(\frac{Q_{b,i}}{RT}\right) c_i = c_{ref,i} + a_j\left(\frac{T}{T_{ref}} - 1\right)$

Zhao et al. found that as the adsorption temperature increased, the selectivity of activated carbon decreased while the selectivity of zeolite 13X increased [9]. The study provides insightful information on configuring the thermogravimetric analysis (TGA) experimental cycles. Moreover, their study shows a comparative analysis of selectivity as a metric for measurement.

In the study conducted by Wei et al., the kinetics of activated carbon desorption were investigated [10]. Activated carbon activated by potassium hydroxide (CA) and activated carbon activated by CO_2 (PA) were the subjects of study for CO_2 capture. TGA was used to investigate CA and PA's adsorption processes. The PA and CA were prepared by waste ion exchange resins. The adsorbents were heated at 393K and saturated in N_2 to remove CO_2, H_2O and impurities from the adsorbents' surface. The experiment incorporated various CO_2 concentration and heating rates to evaluate the kinetic performance and adsorption capacity of the adsorbents. The study demonstrates the approach of utilizing the isothermal thermal analysis to enable the kinetics of CO_2 to be researched and found the kinetics of CO_2 on the adsorbents can be best described by the Avrami-Eroféev model.

From Zhao et al.'s study, zeolite 13x showed to be a promising material in terms of adsorption selectivity for CO_2 [9]. In Guo et al.'s research, the desorption kinetics of CO_2 from zeolite 13X was investigated by TGA analysis [11]. Zeolite 13x was characterized for its pore size distribution (PSD) by the Brunauer-Emmett-Teller (BET) method. The properties of the zeolite 13x were found to consist of 52% total pore volume with micropores (0.7-2 nm) and 48% mesopores (2-50 nm) [11]. To evaluate the kinetic parameters of zeolite 13x, the experimental setup of the TGA comprises two zones, carrier gases, and four heating rates. The TGA and differential thermo-gravimetric (DTG) curves can be seen in Figure 2.

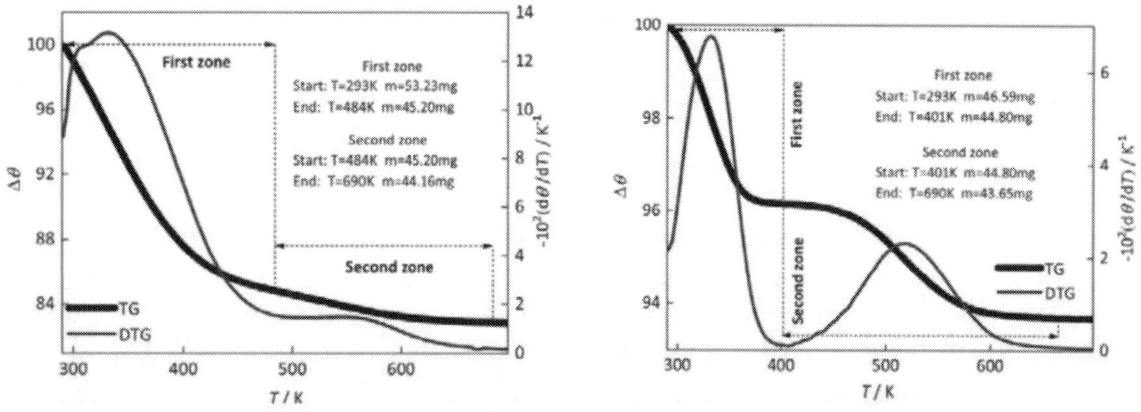

Figure 2. TGA and DTG curves for (a) CO_2 as the carrier gas (b) for argon as the carrier gas [11].

2.2 *Adsorption Capacity*

Adsorption isotherms distinguish the surface chemistry of the adsorbents and provide information such as the adsorbent's capacity and CO_2 uptake at various partial pressures. Figure 3 shows the six types of adsorption isotherms. Type I is the adsorption of gases onto microporous solids such as activated carbon, molecular sieve zeolites, and MOFs. Type II is gas adsorption onto non-porous or microporous solids consisting of condensation phenomena. Type III increases exponentially until partial pressure reaches unity. Type IV is similar to Type I, except it comprises an adsorption uptake at low partial pressure and is typically seen in multilayer adsorptions. Type V includes a gentle uptake at low partial pressures, a rise in uptake as partial pressure increases, and then approaching saturation at higher concentrations. Both Type IV and V correspond to multilayer adsorption followed by capillary condensation; Type VI is an adsorption process consisting of a multilayer non-porous surface [7,8].

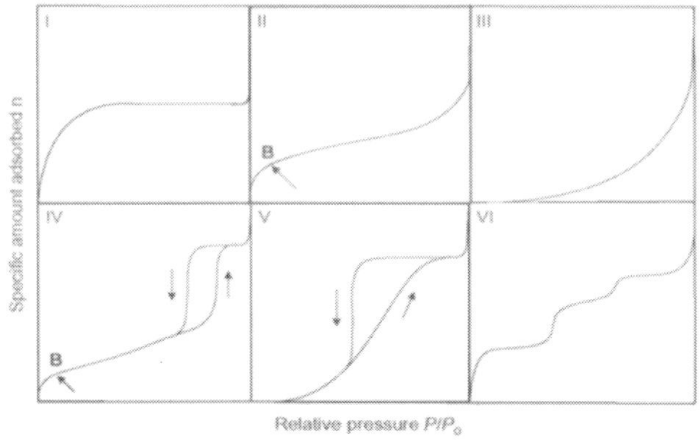

Figure 3. The six types of adsorption isotherm I to VI [7]

In Bezerra et al.'s study, the adsorption capacity of granular activated carbon and zeolite 13X for the capture of atmospheric CO_2 were evaluated by TGA analysis [12]. The adsorption isotherms for activated carbon and zeolite 13X were constructed by assessing the amount of CO_2 adsorbed per gram of sorbent used with respect to the partial pressure, and the results are reported in Table 4. The surface area of the activated carbon and zeolite 13X was 1700 $\frac{m^2}{g}$ and 515 $\frac{m^2}{g}$, respectively [12]. The samples were degassed until no further mass variation was observed, and then the samples were left to cool down to 298 °K. The relative pressure ($\frac{P}{P_0}$) of CO_2 was increased incrementally up to 10 bar, the experimental temperature range was reported to be 298-348 °K, and the amount of CO_2 uptake per gram of adsorbent as a function of $\frac{P}{P_0}$ of CO_2 was studied.

3 Results and Discussion

3.1 Desorption Kinetics

In Wei et al.'s study, the Avrami-Eroféev model was employed to acquire the activation energy. The method comprises of a plot of $ln\left(A\,exp\left(-\frac{E_a}{RT}\right)\right) vs.\frac{1}{T}$ where A is the pre-exponential factor, E_a is the activation energy, R is the gas constant and T is the temperature. E_a can be calculated from the slope of the plot, and the pre-exponential factor (A) can be obtained from the intercept. The activation energy and the pre-exponential attainment allow for the reaction rate constant $K(T)$ calculation. The R-squared (R^2) value denotes how the data fits into the regression model. The results showed $K(T)$ to increase with increasing temperature, which suggests that higher temperatures are advantageous for the desorption of CO_2, and the kinetic parameters are summarized in Table 2 [10].

Table 2. Reaction rate constant $K(T)$, activation energy (E_a), and pre-exponential factor (A) for CO_2 desorption at various temperatures [10]

Sample	Temperature (°K)	K(T)	E_a ($\frac{kJ}{mol}$)	A (min-1)	R^2
Chemically Activated Carbon	373	0.406576	9.4713	8.5584	0.9942
	398	0.48355			
	423	0.58258			
Physically Activated Carbon	373	0.13444	12.6574	7.8428	0.9728
	398	0.16558			
	423	0.21825			

Guo et al.'s found as the heating rate increases, the value of the pre-exponential factor also increases, corresponding to an increase in the desorption rate of zeolite 13X [11]. The desorption kinetics and its corresponding results are shown in Table 3.

Table 3. Desorption kinetics for zeolite 13X [11]

Carrier gas		β (K min^{-1})	E (kJ mol^{-1})	E_{ave} (kJ mol^{-1})	ln A (min^{-1})	n
CO_2	First zone	8	14.12	13.35 (13.35 − 1.49, 13.35 + 1.49)	4.87	1.2
		12	14.07		5.31	1.1
		16	13.04		5.60	1.2
		20	12.15		5.80	1.2
	Second zone	8	58.53	58.24 (58.24 − 0.45, 58.24 + 0.45)	7.25	2.0
		12	58.42		7.44	2.0
		16	57.95		8.26	2.0
		20	58.05		8.37	1.6

Guo et al.'s research confirmed that the adsorption of CO_2 on activated carbon and zeolite is a physisorption process; thus, the activation energy is disregarded, and only the desorption kinetics is sought after [11]. From Wei and Guo et al.'s study, the desorption activation energy is to range from 9 to 14.5 $\frac{kJ}{mol}$ for activated carbon and zeolite [10,11].

3.2 Adsorption Capacity

The research conducted by Bezarra et al. showed CO_2 uptake to increase with increasing amount of CO_2 partial pressure. The amount of adsorbed CO_2 per gram of sorbent used respective to the partial pressure of CO_2 are given in Table 4.

Table 4. Adsorption amount at various temperatures and pressures [12]

Adsorbent	Adsorbed amount $\left(\frac{mg\ CO_2}{g}\right)$					
	298 °K			348 °K		
	0.1 bar	1 bar	10 bar	0.1 bar	1 bar	10 bar
Activated carbon	10	83	340	4	31	156
Zeolite 13X	83	206	255	54	142	197

Bezerra et al. confirmed the adsorption isotherm for activated carbon is Type I isotherm. Extrapolating to a partial pressure of 0.04% for activated carbon and zeolite 13X resulted in an uptake of 1.92 $\frac{mg\ CO_2}{g}$ and 69.388 $\frac{mg\ CO_2}{g}$, respectively. The obtained isotherm showed an increase in uptake until the partial pressure reached unity, indicating that the adsorbent has an excellent pore distribution [12]. The type of isotherm, however, was not confirmed for zeolite 13X but it is ascertained activated carbon to conform to Type 1 isotherm.

4 Conclusion

In this work, the desorption kinetics and adsorption capacity for activated carbon and zeolite were reviewed. TGA analyses provided a comprehensive analysis on the adsorbents' performance and adsorption capacity. For desorption kinetics, activated carbon was found to underperform zeolite. In terms of adsorption capacity, activated carbon outperformed zeolite at settings with higher partial pressure of CO_2. The studies supported higher pressure settings induce the physisorption regime, thus increasing the interaction of CO_2 to polar molecules located at the surface and therefore increasing throughput. The results reported in this work emphasizes the importance of the adsorbents' surface characteristics for estimating and predicting CO_2 adsorption performance.

References:

[1] Baker S.E., Stolaroff J.K., Peridas G., Pang S.H., Goldstein H.M., Lucci F.R., Li W., Slessarev E.W., Pett-Ridge J., Ryerson F.J., Wagoner J.L., Kirkendall W., Aines R.D, Sanchez D.L., Cabiyo B., Baker J., McCoy S., Uden S., Runnebaum R., Wilcox J., Psarras P.C, Pilorgé H., McQueen N., Maynard D., McCormick C., *Getting to Neutral: Options for Negative Carbon Emissions in California*, January, 2020, Lawrence Livermore National Laboratory, LLNL-TR-796100

[2] Sandalow D., Friedmann J., McCormick C., and McCoy S (2018). *Direct Air Capture of Carbon Dioxide.* Innovation for Cool Earth Forum.

[3] Gregg S.J. and Sing KSW (1982). *Adsorption, Surface Area and Porosity.* Academic Press Inc. (London) LTD.

[4] Estevez, L., Barpaga, D., Zheng, J., Sabale, S., Patel, R. L., Zhang, J. G., McGrail, B. P., & Motkuri, R. K. (2018). Hierarchically Porous Carbon Materials for CO2 Capture: The Role of Pore Structure. *Industrial and Engineering Chemistry Research*, *57*(4), 1262–1268. https://doi.org/10.1021/acs.iecr.7b03879

[5] Wang, Y., Wang, M., Wang, Z., Wang, S., & Fu, J. (2020). Tunable-quaternary (N, S, O, P)-doped porous carbon microspheres with ultramicropores for CO2 capture. *Applied Surface Science*, *507*. https://doi.org/10.1016/j.apsusc.2019.145130

[6] Surampalli, R. Y., Brar, S. K., Gurjar, B. R., Kao, C., Naidu, R., Ojha, C. S. P., Ramakrishnan, A., Tyagi, R. D., Yan, S., & Zhang, T. C. (Eds.). (2014). *Carbon capture and storage : Physical, chemical, and biological methods*. American Society of Civil Engineers.

[7] Erkey C. (2011). *Thermodynamics and Dynamics of Adsorption of Metal Complexes on Surfaces from Supercritical Solutions.* https://www.oreilly.com/library/view/supercritical-fluids-and/9780080914480/OEBPS/B9780080453293000044.html .

[8] Nix R. (2021). PE Curves and Energetics of Adsorption. Chemistry Libre Text. https://chem.libretexts.org/Bookshelves/Physical_and_Theoretical_Chemistry_Textbook_Maps/Book%3A_Surface_Science_(Nix)/02%3A_Adsorption_of_Molecules_on_Surfaces/2.04%3A_PE_Curves_and_Energetics_of_Adsorption.

[9] Zhao R., Liu L., Zhao L., Deng S., Li S., and Zhang Y. (2019). *A Comprehensive Performance Evaluation of Temperature Swing Adsorption for Post-Combustion Carbon Dioxide Capture.* Renewable and Sustainable Energy Reviews 114 (2019) 109285.

[10] Wei M., Yu Q., Xie H., Zuo Z., Hoi L., and Yang F. (2017). *Kinetic Studies of CO_2 Adsorption and Desorption on Waste Ion-Exchange Resin-based Activated Carbon.* International Journal of Hydrogen Energy 42 (2017) 27122-27129.

[11] Guo Y., Zhang H., and Liu Y. (2018). *Desorption Characteristics and Kinetic Parameter Determination of Molecular Sieve by Thermogravimetric Analysis/ Differential Thermogravimetric Analysis Technique.* Adsorption Science & Technology 2018, Vol. 36 (7-8) 1389-1404.

[12] Bezerra D.P., Oliveira R.S., Vieira R.S., Cavalcante Jr. C.L., Azevedo D.C.S. (2011). *Adsorption of CO_2 on Nitrogen-enriched Activated Carbon and Zeolite 13x.* Adsorption (2011) 17: 235-246 DOI 10.1007/s10450-011-9320-z.

639428 Hydrophobic Natural Based Deep Eutectic Solvents for High Performance Carbon Dioxide Solubility

Abstract:

Natural deep eutectic solvents (NADES) were studied in this work on their carbon dioxide (CO2) capture performances. Monoterpenoids (cineole, carvone) were used as hydrogen bond acceptors (HBA), and they were mixed with other terpenes (menthol, thymol). The production of NADES systems was carried under specific conditions of mixing time, mixing speed, temperature, and molar ratio between the HBD and the HBA. These conditions were specified to reach the best point of eutectic between the HBD and the HBA. The optimized molar mixing ratios of the hydrogen bond acceptor and hydrogen bond donor have been identified by COSMO-RS calculations along with the sigma profiles for each eutectic system. Water content on these materials measured via Karl-Fisher coulometric titrator to confirm their hydrophobic behavior. Density, viscosity, conductivity, pH, surface tension and contact angle (on glass, stainless steel, copper) characterization at different isotherms have been carried out prior to experimental gas solubility tests. The carbon dioxide solubility experiments were performed in a high-pressure isochoric equilibrium autoclave. The carbon dioxide solubility experiments were conducted up to 40 bars at two isotherms (25°C and 35°C). The solubility data was benchmarked against the other state of the art solvents and emerging ionic liquid based solvents. The COSMO-RS solubility predictions were also used to compare the experimental findings against software predictions. The data has shown high solubility at low temperature isotherm around 40 bars. Carvone and Thymol based NADES has shown the highest solubility performance amongst the other studied NADES systems.

ADMETSTAR 2.0 toxicity calculations showed non-toxic behaviour on these NADES making them suitable agents for greener applications. Considering their low viscosity and high carbon dioxide solubility, these solvents and their other combinations with monoterpenoids can be considered for future applications in industry in large scales.

639429 A Nanoscopic Explanation on Hydrophobic Deep Eutectic Solvents and Their Carbon Dioxide Solubility Performance

Abstract:

Type V Natural Deep Eutectic Solvents (NADES) considering carvone, 1,8-cineole, menthol, thymol are studied considering a theoretical approach with the purpose of developing a multiscale characterization of these fluids with particular attention to intermolecular forces (hydrogen bonding) of NADES and carbon dioxide, and their relationships with macroscopic behavior. Supporting experimental data has also been collected during this study and compared to theoretical data. Density functional theory (DFT) calculations with at B3LYP/6-311++G** theory levels and we have shown that this theory/basis-set combination works very well on such systems. Quantum Theory of Atoms in a Molecule (QTAIM, Bader's theory) using MultiWFN code has been applied for the analysis of bonding structures and strengths between the hydrogen bond acceptor and hydrogen bond donor of the NADES between the carbon dioxide molecule. From this analysis, interaction regions are characterized by the developed bond critical points and the corresponding values of electron density, ρ, and the Laplacian (∇) of electron density, $\nabla^2\rho$. Classical molecular dynamics simulations were also performed allowing to characterize bulk liquid phases at the nanoscopic level, analyzing the fluid's structuring, voids distribution and dynamics. The reported results allowed to infer nano – macro relationships, which are required for the proper design of these green solvents and their application for different technologies. It has been shown that the suitable interactions of NADES with CO_2 did not result disruptions in the solvent hydrogen bonding network. Moreover, predicted ecotoxicological properties have showed low impact fluids, which may be considered as suitable alternatives for different operations including carbon capture.

SPRING22
+18TH GCPS
A Joint AIChE and CCPS Meeting

Carbonomics: Introduction to Carbon Pricing, Regulations, and Frameworks

Michael C. Stern
Exponent, Inc.
1075 Worcester Street, Natick, MA 01760
mstern@exponent.com

Jason James,[a] Ann Michelle Morrison,[b] Achim Wechsung,[c] Christopher Buehler[d]
Exponent, Inc.
a - 15375 SE 30th Place, Suite 250, Bellevue, Washington 98007
b – 1 Mill and Main Place, Suite 150, Maynard, Massachusetts 01754
c – 1075 Worcester Street, Natick Massachusetts 01760
d – 420 Lexington Avenue, Suite 1740, New York, New York 10170

Prepared for Presentation at
American Institute of Chemical Engineers
2022 Spring Meeting and 18th Global Congress on Process Safety
San Antonio, TX
April 10 – April 14, 2022

Carbonomics: Introduction to Carbon Pricing, Regulations, and Frameworks

Michael C. Stern
Exponent, Inc.
1075 Worcester Street, Natick, MA 01760
mstern@exponent.com

Jason James, Ann Michelle Morrison, Achim Wechsung, Christopher Buehler
Exponent, Inc.

Keywords: Carbon Capture & Storage, Carbon Management and Economics & Finance

Abstract

Chemical engineers developing sustainable technologies or managing sustainability programs should be aware of the economics, regulations, and frameworks surrounding carbon. Carbon economics, regulations, and frameworks include mandatory cap-and-trade markets, voluntary markets, border adjustment mechanisms, science-based emissions targets based on the Paris Accord, and the cost of a range of carbon avoidance and sequestration-based offsets such as reforestation, soil sequestration, and direct air capture. Further, engineers should be aware of emerging frameworks from a range of organizations and federal agencies including the evolving disclosure requirements being developed by the Securities and Exchange Commission (SEC), and reporting metrics developed by the Taskforce on Climate-Related Financial Disclosures (TCFD).

1 Introduction to Carbonomics

Carbonomics, while not a strictly defined term, refers the economic consideration around the avoidance, generation, release, capture, sequestration, and utilization of carbon. However, carbonomics cannot be sufficiently addressed by any single economic, scientific, or engineering discipline because it is based on the complex interconnections between these disciplines. This creates a challenge for companies to create a methodology to price and document carbon emissions and reductions in a field where multiple disparate groups have differing nomenclature and accounting methods.

Creating equivalencies between widely varying techniques aimed to achieve the same objective is one such challenge. For carbon sequestration, carbon dioxide (CO_2) can be injected into deep wells where it would presumably remain for hundreds if not thousands of years. At the same time, carbon can also be absorbed in forests, mangroves, and soils where it might stay for a hundred years (or just a few years in the event of a fire). Is the value of both types of sequestrations equivalent? If not, how does one price the value of longer duration sequestration or account for the risk of loss back to the atmosphere through leakage and wildfire?

In reality, the answers to these questions are determined by the governing regulations and frameworks that surround the accounting practices. These regulations and frameworks are ideally informed by scientific considerations, but they can be difficult to adapt to the rapidly changing scientific understanding of what methods and technologies are most effective.

2 Carbon Regulations and Frameworks

CARB Compliance Offset Program

One of the most used systems to determine carbon offsets is based on regulations promulgated by the California Air Resource Board (CARB).[1] This system includes mechanisms to grant offset credits for forestry, livestock, mine methane capture, and ozone depleting substance projects. CARB also has designations for "Urban Forest" and "Rice Cultivation" projects but has yet to issue any offset credits for them.[2] However, CARB has received criticism for the way they issue forestry offsets, with claims that the issued credits overestimate the benefits of the forestry work and underestimate the risk of wildfires to forestry projects.[3]

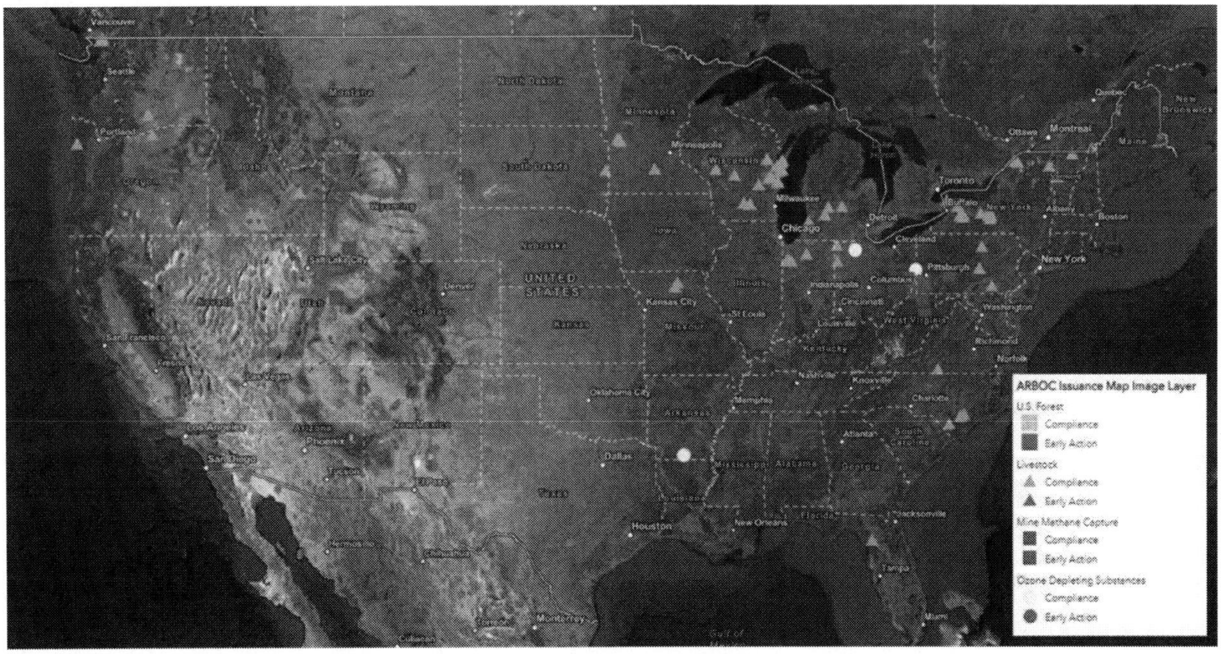

Figure 1. Geospatial map of CARB Compliance Offset locations (available at https://webmaps.arb.ca.gov/ARBOCIssuanceMap/).

Non-Governmental Certification

A number of non-governmental entities have created voluntary markets for carbon offsets. These offsets are verified based on standards developed by non-governmental organizations and are not subject to specific government oversight. The UN Clean Development Mechanism (CDM), which was established as part of the Kyoto Protocol, was the first major mechanism by which carbon offset credits could be traded.[4] The most widely used voluntary greenhouse gas (GHG) verification program is currently developed by Verra.[5]

Companies, law firms, and other brokers then work as intermediaries to help various interested parties purchase voluntary credits from establishments developing projects that either sequester or avoid emissions.

Interested parties purchasing voluntary credits may want to carefully vet the projects from which they buy credits to ensure they feel comfortable with the quality of the project, the type of offset credit (sequestration vs. avoidance), and the potential risks of loss due to leakage from subsurface injection[6] or wildfire. Some stakeholders may find certain carbon offset projects may have reputational or legal risks if the source of the credits is publicly questioned.

Non-Governmental Reporting Frameworks

GHG emitters and purchasers of carbon offset credits should understand how their emissions will be reported to the public. A number of non-governmental frameworks exist including the Taskforce on Climate-Related Financial Disclosures (TCFD) and the GHG Protocol.[7,8] These two frameworks were relied upon significantly by the Securities and Exchange Commission (SEC) proposed rules released in March 2022.[9] Other entities such as the Sustainable Accounting Standards Boards (SASB) and the Climate Disclosure Project (CDP) offer additional industry specific disclosure guidance and a disclosure platform, respectively.[10,11]

These reporting frameworks typically require metrics that include total absolute emissions and emissions intensity (emissions per amount of product) as well as more subjective information such as climate risks for businesses. For example, companies that use natural resources may need to disclose their perceived risks for the possible loss of those resources due to climate change.

Science-Based Targets (SBTs)

Science-Based Targets (SBTs), administered through the Science-Based Targets initiative (SBTi), is a collaboration between CDP, the UN Global Compact, the World Resources Institute, and the World Wide Fund for Nature. SBTs are considered "science-based" if they are in line with what the latest climate science deems necessary to meet the goals of the Paris Agreement – limiting global warning to well-below 2°C above pre-industrial levels and pursuing efforts to limit warming to 1.5°C.[12]

It is important to note that the primary determinant of an SBT is that it is a sufficient emissions reduction target to be in alignment with what climate science deems necessary. Therefore, it is possible for companies to make SBTs that may not yet be technologically or economically feasible. Because of the timeline of SBTs, which span decades, it would be unrealistic to assume that existing technology and methods would be wholly sufficient to reach goals for GHG reductions by 2050. Further, carbon offsets are not accepted as equivalent to reductions in carbon emissions, requiring a technical (rather than financial) mechanism to meet SBTs.[13] Therefore, engineers and scientists will not only be part of shaping corporate SBTs but also necessary to ensuring companies can meet those targets.

3 Carbon Pricing

Based on the above frameworks, carbon offsets have become tradeable commodities.[14] However, the market remains complicated as there are still multiple types of carbon offsets based on different frameworks and regulations. As of the time of this paper, dozens of different frameworks exist with carbon prices ranging from over \$100/ton CO_2 equivalent (tCO_2e) in Switzerland and Sweden to just a few dollars or less per tCO_2e in range of locations from Massachusetts to eastern Europe to Asia.

Engineers performing initial calculations to assess the profitability of new facilities that either will require or create carbon offset credits may need to consider the location, industry, stakeholder engagement, and duration of the project to estimate the value or cost of the associated credits. For projects with decades long time horizons, such estimates will be subject to significant uncertainty.

4 Conclusions

Chemical engineers have traditionally been taught about process economics in terms of feedstocks, energy costs, capital expenses, and operation and maintenance. Going forward, chemical engineers will need to broaden the scope of their economic assessments to include the cost of greenhouse gas emissions or the value of carbon reduction. Such considerations, however, are significantly complicated by the rapidly evolving landscape of carbonomics. The cost (or value) of carbon is poised to increase over the coming years, but whether carbon increases to \$20, \$30, or \$100 per ton CO_2e will likely change the result of many process economics assessments. Engineers involved in process conceptualization or process design should consider not just the current cost of carbon, but also the uncertainty in future cost of carbon, in their assessments.

5 References

[1] California Air Resource Board. Compliance Offset Program. https://ww2.arb.ca.gov/our-work/programs/compliance-offset-program. Accessed April 2, 2022.

[2] California Air Resource Board Issuance Table, https://ww3.arb.ca.gov/cc/capandtrade/offsets/issuance/arboc_issuance.xlsx. Accessed April 2, 2022.

[3] Badgley G, Freemen J, Hamman JJ, Haya B, Trugman AT, Anderegg WRL, Cullenward D. Systematic over-crediting in California's forest carbon offsets program. *Global Change Biology* 2021; **28**: 1433-1445.

[4] United Nations Framework Convention on Climate Change, The Clean Development Mechanism. https://unfccc.int/process-and-meetings/the-kyoto-protocol/mechanisms-under-the-kyoto-protocol/the-clean-development-mechanism. Accessed April 2, 2022.

[5] Verra. Verified Carbon Standard. https://verra.org/project/vcs-program/. Accessed April 2, 2022.

[6] Vinca, et al., "Bearing the Cost of Stored Carbon Leakage." Frontiers in Energy Research, May 2018. Available at https://doi.org/10.3389/fenrg.2018.00040.

[7] Financial Stability Board. Task Force on Climate-related Financial Disclosures. https://www.fsb-tcfd.org/. Accessed April 2, 2022.

[8] Greenhouse Gas Protocol. https://ghgprotocol.org/. Accessed April 2, 2022.

[9] US Securities and Exchange Commission. SEC Proposes Rules to Enhance and Standardize Climate-Related Disclosures for Investors. https://www.sec.gov/news/press-release/2022-46. Accessed April 2, 2022.

[10] Sustainability Accounting Standard Board. https://www.sasb.org/. Accessed April 2, 2022.

[11] Climate Disclosure Project (CDP). https://www.cdp.net/en. Accessed April 2, 2022.

[12] Science Based Targets. https://sciencebasedtargets.org/. Accessed April 2, 2022.

[13] Science Based Targets. SBTi Criteria and Recommendations TWG-INF-002, Version 5.0. October 2021. https://sciencebasedtargets.org/resources/files/SBTi-criteria.pdf. Accessed April 2, 2022.

[14] The World Bank. Carbon Pricing Dashboard. https://carbonpricingdashboard.worldbank.org/. Accessed April 2, 2022.

638900 Restaurant Food Waste in UAE: Life Cycle Assessment (LCA) and Feasibility of Conversion into Usable Biogas through Anaerobic Digestion

Abstract:

Restaurants contribute 32% of UAE food waste generated annually that ends up mostly in landfills. The associated environmental impacts of food waste landfilling are of concern globally. Consequently, alternative waste management technologies have been considered to replace landfill. This study employs life cycle assessment (LCA) to compare landfilling (BaU) of restaurant food waste with two other scenarios namely, anaerobic digestion (AD) and composting (COMP) to investigate the most environmentally beneficial system. The assessment allocated useful byproducts such as electricity and compost as environmental benefits and assessed how those influence the results.

The results showed that anaerobic digestion (AD) performed better in 5 of 6 impact categories assessed – climate change, freshwater eutrophication, terrestrial human toxicity cancer and non-cancer, and fossil depletion. Composting (COMP) showed a lower environmental impact compared to landfilling in 4 of 6 impact categories. The study found that AD showed a significant net environmental benefit especially in the climate change and fossil depletion categories; −1080 kg CO_2 eq. and −459 oil eq. respectively. The environmental impacts of all the assessed systems are influenced by distance traveled to convey waste to the treatment facilities, thus a strategic siting of anaerobic digestion systems to reduce transportation will further improve its environmental performance as a viable food waste disposal system.

Biogas is the main product from food waste anaerobic digestion, which is a renewable resource that can be used for heating, cocking and electricity production. However, to ensure continuous and feasible production of biogas, restaurant food waste is mixed with cattle manure before being processed by 5 anaerobic digesters in series. Such configuration is expected to treat between 1,500 to 2,200 metric tonnes of restaurant food waste, which is approximately 0.05% of Dubai's annual food waste based on 2015 figures.

Such AD configuration is expected to produce annually between 0.29 to 0.41 *million cubic meters of biogas* with expected unit cost of $1.86 – 2.67 per *cubic meter. The biogas* can be sold directly as cooking gas, or it can be converted into electricity to cover the daily consumption of 130 – 190 persons.

The outcome of the study can be used as a basis by the UAE government to achieve its goal of diverting 75% of municipality solid waste from landfills. Also, the study shows that anaerobic digestion plants integrated with CHP units can be strategically sited across the country to generate clean electricity thereby reducing the environmental impact of transportation. However, further study is required to assess the financial and economic feasibility of suitable AD system for the UAE environment.

639598 Impact of COVID-19 Pandemic on Petrochemical Industry Emissions

Abstract:

As a result of the COVID-19 and subsequent restrictions in spring 2020, enormous lockdown orders were issued to prevent the spread of the pandemic. All those restrictions led to substantial interruptions in both individuals' daily life and businesses activities worldwide. Texas, which was not an exception, received stay-at-home order from April to the beginning of May 2020. To quantify the temporal impact of COVID-19 pandemic on petrochemical industries, emissions concentration fluctuations during the pandemic in Jefferson County, Texas has been studied in this work. First, various emissions in Jefferson County (including Beaumont and Port Arthur, twin cities which are home to many oil and gas companies) were tracked individually. The emissions concentrations in April 2020 were compared to those in April 2019 based on statistical methods. Second, the emission concentrations in pre-lockdown time period (January, February, and March 2020) were compared to the data set of the same three months in 2019. The public data of volatile organic compounds (VOCs), NO_2, and SO_2 was used to track the emissions changes in the lockdown. Based on our obtained results, the emission quantities in Jefferson County, Beaumont, and Port Arthur had significantly dropped by 45.71%, 54.43%, and 58.82% in April 2020 in comparison to April 2019 respectively. Alkanes, acetylene 1,3- butadiene, ethylene, benzene, and toluene were major VOC compounds in reduction. Additionally, the result showed emissions reduction in the lockdown time period was exceptional compared to January, February, and March 2020.

640190 Wind Impact on Air-Assisted Flare Performance

Abstract:

In this work, computational fluid dynamics (CFD) was used to predicate the impact of wind speed on the air-assisted flare performance with a new design tip. CFD is a powerful and reliable technique that can be used to investigate the flaring process under different scenarios, which can be costly and risky. The simulation model was first developed and validated against the available experimental data, then the model was used to simulate all simulation cases. Three different wind speeds of 5, 10, and 15 m/s were used for the same waste gas flow rate of 0.5 ft/sec. Propane gas was used as a flaring gas that flowed through a 6-inch flare. The results show that the flare performance at high wind speeds caused a decline in the flare performance when the new tip was not involved in the operation. The results also showed that the impact of the high wind speed on the flare performance was significantly declined when the air was injected from the new tip. The new tip injected air has a significant impact on the flare performance when the flow of waste gases is very low, and the wind speed is high.

640103 Synthesis of Graphene Based Iron-Copper Nano-Composites and Its Application in Removal of Methylene Blue

Abstract:

Herein, we report the in-situ synthesis of FeCu nanoparticles with varying concentration on the reduced graphene oxide using $NaBH_4$ as a reducing agent. The bimetallic-reduced graphene oxide nanocomposite (rGO-FeCu) is characterized by transmission electron microscopy (TEM), and X-ray diffraction (XRD). TEM revealed that average 6nm particles were decorated on graphene sheet. To the best of our knowledge, this is the first report on FeCu bimetallic decorated on the rGO in the degradation of methylene blue (MB) dye. The system of rGO-FeCu and per-carbonate (PC) was used for degradation of dye, which is a highly toxic contaminant in water. The removal of dye was investigated by using rGO-FeCu in the absence and presence of PC. Furthermore, PC concentration, catalyst loading, and dye concentration were systematically studied to understand their effects on the degradation of MB.

640111 Synthesis of Graphene Based Ferrites and Its Application in Removal of Heavy Metals

Abstract:

$ZnCeFe_2O_4$ based ferrites supported on the graphene oxide (GO-$ZnCeFe_2O_4$) are synthesized through sol-gel method which exhibit enhanced efficiency in the removal of mercury. The nanocomposite is characterized by Scanning electron microscopy (SEM), FTIR and XRD. Nano-sized GO-$ZnCeFe_2O_4$ are successfully synthesized as confirm by SEM and XRD. To better understand removal efficiency of GO-$ZnCeFe_2O_4$, we conducted the heavy metal removal activity of different composition of nanocomposite for mercury. In the same concentration and time conditions, graphene based ferrites shows the highest removal efficiency, as compare to the others. Chemical Oxygen Demand (COD) tests were also conducted to verify the catalyst results. This GO-$ZnCeFe_2O_4$ represent a highly potential material for heavy metals removal and remediation of other contaminants in the wastewater.

SPRING22
+18TH GCPS
A Joint AIChE and CCPS Meeting

Process Solutions for Improving the Stability and Handling Properties of Pyrolysis Oils

Daniel Dreyer, Kameswara Vyakaranam, Kuldeep Wadhwa, Dave Smith
Nalco Water, An Ecolab Company
11177 S. Stadium Dr., Sugar Land, TX, 77478
daniel.dreyer@ecolab.com

Prepared for Presentation at
American Institute of Chemical Engineers
2022 Spring Meeting and 18th Global Congress on Process Safety
San Antonio, TX
April 10 – April 14, 2022

Process Solutions for Improving the Stability and Handling Properties of Pyrolysis Oils

Daniel Dreyer, Kameswara Vyakarnam, Kuldeep Wadhwa, Dave Smith
Nalco Water, An Ecolab Company
11177 S. Stadium Dr., Sugar Land, TX, 77478
daniel.dreyer@ecolab.com

Keywords: chemical recycling, pyrolysis oil, circular economy

Abstract

As governmental regulations and a variety of public and private companies shift toward the realization of a circular economy, chemical recycling technology is seeing increased interest as a key strategy in managing plastic waste. Polypropylene (PP) and polyethylene (PE) represent the largest volumes of plastic waste, and thus most chemical recyclers are focused on converting PP and PE into what is known as pyrolysis oil. Pyrolysis oil can then be used as a feedstock for a variety of processes, including steam crackers that are used to produce ethylene used in the production of virgin PE resins, thus completing the full lifecycle of polyolefin use, waste, and reuse.

Key areas of concern have been identified, including cold flow handling behavior and stability (particularly with regards to the formation of sediment, gums, and/or films). Treatment with various chemical additives provides customers with options for addressing these challenges and improving the overall properties and performance of the pyrolysis oil stream.

Here we provide a discussion of broad industry trends around chemical recycling, initial findings regarding the challenges associated with pyrolysis oils, and chemical treatment options that can be considered and have been demonstrated at lab scale.

1 Introduction

While plastics have many beneficial aspects – from safe packaging for food and water, to construction materials, medical devices, and transportation – the negative impacts of plastic waste are resulting in consumers and governments demanding more sustainable alternatives to single-use plastics. For example, the European Union has banned certain single-use plastics and has enacted taxes on other non-recyclable materials,[1] as have China, Canada, and American states like California.[2-4] Moreover, consumers are demanding more sustainable products, and corporations are setting aggressive sustainability and climate change-related goals.

An estimated 350 million metric tons of plastics are produced globally each year, with demand forecasted to increase in the coming decades.[5] Currently, only approximately 10% of plastic waste is reused or recycled, with the remainder ending up in incinerators, discharged to landfills, or discarded into the environment. According to the United Nations, the equivalent volume of a garbage truck of plastic is added to the ocean every minute, and without action, by 2050 the ocean will contain more plastic than fish.[6]

Traditional recycling processes have generally relied on mechanical methods. The plastic is subjected to shear forces, such as shredding, and then melted, compounded, and extruded for reuse. Mechanical recycling inevitably results in the physical degradation of the polymer material, meaning that with each successive recycle, the material is downgraded in value. Moreover, with successive recycles, contaminants such as pigments, dyes, and other additives can accumulate in the recycled material.

In the past few years, however, new chemical recycling methods have been developed that are truly circular in nature – that is, chemical processes are used to degrade the polymeric chains, thereby affording high value monomer and feedstock for the continued circular production of plastic. The processes are molecular rather than physical, so the product is not downgraded by recycling as it is in mechanical recycling.

Table 1. Representative methods for chemical recycling.

Monomer Recycling	**Plastic-to-Feedstock (PTF) Recycling**
Chemolysis	Pyrolysis
Hydrolysis	Fluid catalytic cracking
Methanolysis	Hydrogenation
Glycolysis	Catalytic pressure conversion
Aminolysis	Gasification
Other methods	Hydrothermal liquefaction (supercritical water)
	Other methods

There are multiple chemical recycling processes, but they can be characterized into two general groups (Table 1): monomer recycling (also known as depolymerization), and plastics-to-feedstock (PTF) recycling (also known as feedstock recycling or plastics-to-liquid recycling). In monomer recycling, a polymer chain is broken down to well-defined molecules, often monomers. For example, polystyrene can be thermolyzed to produce styrene monomer (Figure 1), which can then be reprocessed to make virgin polystyrene.

Figure 1. Thermoylsis of polystyrene to produce virgin styrene monomer.

In PTF recycling, one of a variety of pyrolysis technologies can be used to crack the polymer chains into various products that can then be used as fuel, or for true circularity, feedstocks for ethylene steam crackers or FCC units in refineries (Figure 2).[7] The pyrolysis oils that are generated via this approach are neither plastic nor monomer and have a unique set of chemical and physical properties.

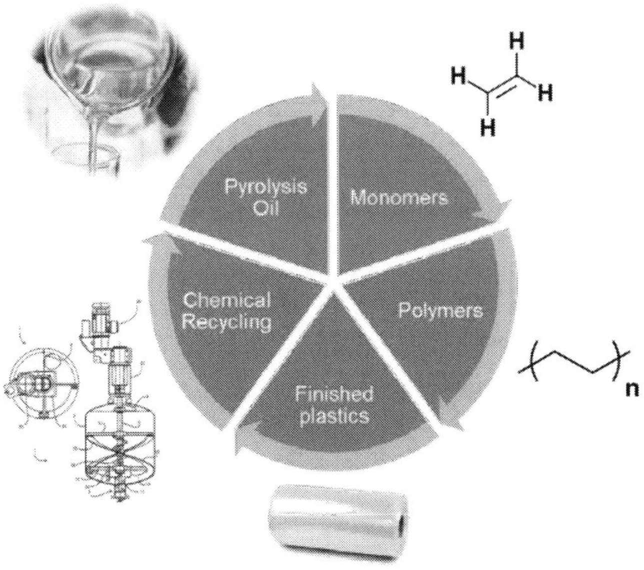

Figure 2. Plastics circularity via chemical recycling.

The response of the chemical processing industry to chemical recycling and the idea of circulatory in manufacturing has been swift, sustained, and widespread. Industry groups such as the Alliance to End Plastic Waste, consisting of over 90 member companies such as Dow Chemical, ExxonMobil, BASF, and others, have committed billions of dollars to projects around the world to focus on reducing waste through various projects.[8] Chemical manufacturers have formed partnerships with chemical recyclers and technology licensors to launch new recycling endeavors and expand the scope and scale of those already in place.

While currently most chemical recycling facilities are operating at sub-commercial scale, companies and partnerships are investing capital to scale up production capabilities,[9] and chemical manufacturers are setting aggressive targets for back-integration, with the ultimate goal of having facilities that can produce circular feedstock onsite with the ethylene manufacturing facility or FCC.[10] Companies such as Chevron Phillips Chemical (CPChem) are already taking advantage of circularly sourced feedstocks – CPChem recently announced the addition of a circularly-produced polyethylene brand called Marlex Anew.[11]

Processing recycled plastics can bring challenges, however – to the chemical recycler themselves, to the chemical industries and refineries seeking to use circular feedstock, and to the technology licensors. The purpose of this paper is to detail the research that Nalco Water has put into identifying these problems and developing solutions.

2 Challenges with Pyrolysis Oils

While the chemical recycling of PE and PP presents remarkable commercial opportunities, and has been studied for at least 50 years,[12] the production and processing of pyrolysis oil streams has brought to light an array of new challenges. Much of this work has focused on identifying and increasing the scale of pyrolysis technology (e.g., reactor designs, reaction conditions, catalysts, etc.). In addition to the challenges of the pyrolysis process itself, however, the streams produced have a wide range of issues associated with stability and handling. Nalco Water is committed to understanding these problems and working with customers to develop solutions. We have found that these issues are manageable, as will be discussed below, but should be acknowledged at the outset.

For example, in the absence of suitable processing equipment (e.g., hydrotreatment), the pyrolysis oil that is produced contains a high degree of unsaturation. Olefinic species, including highly reactive diolefins, are an unavoidable product as high MW polyolefins are degraded into progressively smaller chains, and these species are widely known to be oxidatively unstable. The formation of gums or sediment, discoloration and fouling can all occur, resulting in pyrolysis oils with unexpected properties and/or impairment of process equipment.

Additionally, over the course of our investigation of various pyrolysis oil samples, we have noted that films and sediment can form even if a sample is kept free of oxygen or treated with high dosages of antioxidant. Thus, there appears to be an alternate route for solids formation and fouling. We believe this process is related to impurities present in the feedstock (for example due to poor sorting or contamination of the waste plastic stream), which may be present at widely varying levels depending on the composition of the plastic supply.[13] Given that this fouling process appears to progress regardless of whether an oxidizing atmosphere is present, management strategies other than antioxidant treatment are likely necessary.

A final area of challenge worth highlighting is related to the cold flow properties of these pyrolysis oils. The composition of the finished streams can vary significantly depending on the waste plastic stream and the pyrolysis conditions used,[14] but in many cases (particularly if the pyrolysis conditions are mild or if the product is being fractionated after pyrolysis) the oil can have a waxy consistency. Handling and transporting waxy streams like this (whether derived from plastic recycling or fossil fuels) can be challenging, and suitable additization is often needed to avoid process disruptions.

3 Additive Solutions

While some of the challenges associated with handling pyrolysis oil streams can be managed mechanically or operationally (for example, via expeditious use of an oxidatively unstable stream), we have spent over two years developing a suite of products that are designed to allow for an additive-based management approach. We have investigated over 30 different pyrolysis oil streams in our laboratory, and in the course of this work we have identified three separate additives that can be used to treat unstable or otherwise challenging pyrolysis oils.

3.1 Antioxidants

The first of these products is an antioxidant. As noted above, pyrolysis of PE, PP, and other types of mixed plastic waste results in liquid streams with a high degree of unsaturation. Upon exposure to elevated temperatures, or after prolonged storage periods, gums formation, sediment formation and discoloration become serious concerns (Figure 3, left).

Oxidation can be a relatively slow process under normal conditions, and so we turned to accelerated tests that are better suited for laboratory screening. One method that has proven useful is a variation of the Rancimat test. Originally developed for characterizing the oxidative stability of biodiesel and the feedstocks used to produce biodiesel, we have found that modification of the Rancimat test conditions allows rapid and quantitative characterization of pyrolysis oil oxidation stability. That stability is expressed as an induction time, measured in hours, and as shown in Figure 3 (right), induction times can vary substantially depending on the specific pyrolysis oil. Higher induction times indicate a higher level of oxidative stability. Some samples, such as Pyrolysis Oil A or Pyrolysis Oil B, have moderately elevated baseline induction times, and in both cases the induction times increased significantly upon treatment with an antioxidant. Those results indicate that these oils would likely be similarly stabilized in field storage and use conditions upon treatment with antioxidant. Conversely, Pyrolysis Oil C exhibited a much lower baseline induction time and proved to be relatively resistant to stabilization with antioxidant. These results suggest that this stream would be similarly difficult to handle in field conditions. Higher dosages of antioxidant may provide desired levels of stability, but the costs of such a treatment program would need to be considered to determine if that route would be economical.

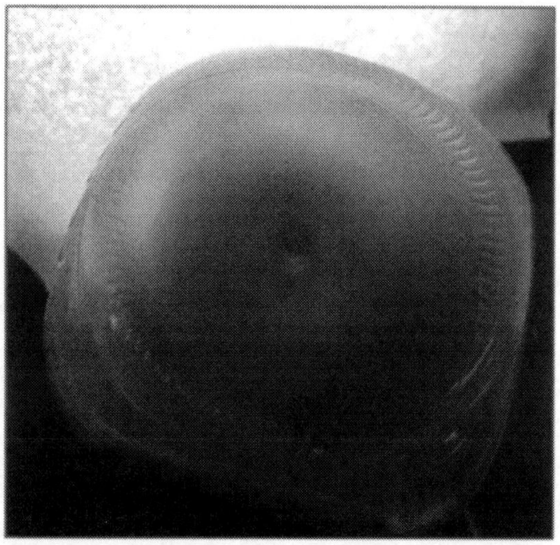

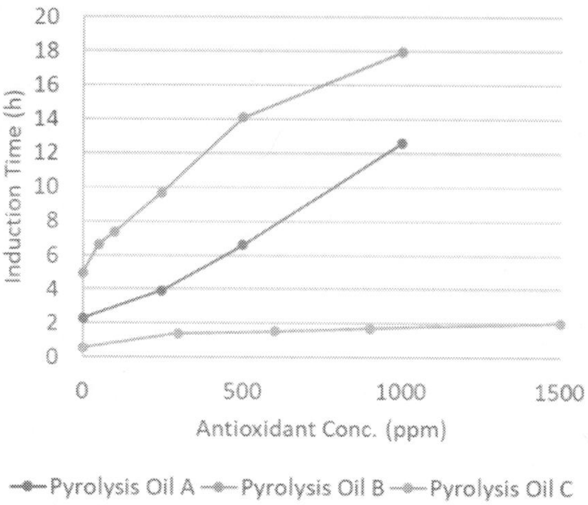

Figure 3. Fouling observed in certain samples of pyrolysis oil (left) and lab-based Rancimat test results indicating improved stability upon treatment with antioxidant (right).

The improved stability that we have seen in our laboratory testing upon treatment with our antioxidant has also been observed in customer field trials. Dosage requirements vary depending on the baseline oxidative stability of the pyrolysis oil, but stability can usually be significantly improved by the addition of 100-1000 ppm of product.

3.2 Extraction Solvents

Non-oxidative film formation is also a challenge that can be addressed using an appropriate chemical treatment program. The second offering we have identified for use in plastics recycling is an extractant whose role is to remove these film-forming species from the stream. The extractant is designed to phase separate from the pyrolysis oil and can be separated via decantation, liquid-liquid extraction or other suitable techniques. As shown in Figure 4, film formation in an untreated sample of pyrolysis oil left a persistent, adherent residue on the inside surface of a bottle held at ambient lab conditions (left). Conversely, when the pyrolysis oil was extracted prior to storage, no signs of film formation or residue were evident (right). We anticipate that this non-oxidative film formation could be problematic for both producers and users of pyrolysis oil streams, and treatment with the extractant is a viable option for mitigating this challenge. Our standard treatment is done with 2 volume percent of the extraction solvent, relative to the volume of pyrolysis oil, but more or less solvent may be necessary depending on how prone a given oil sample is to film formation. In very severe cases, serial treatment may be needed to fully suppress deposition.

Untreated Pyrolysis Oil D　　　　Extracted Pyrolysis Oil D

Figure 4. Untreated (left) and treated (right) samples of pyrolysis oil using an extractant.

3.3 Cold Flow Improvers

The final offering that Nalco Water currently has available for use in improving the properties of pyrolysis oil streams is a pour point depressant (PPD). Depending on how extensively the polymers are degraded during the chemical recycling process, the carbon chain length in the oil can vary substantially. Many of the streams we have tested have relatively high pour points. Such streams have the potential to be problematic, particularly in cold climate environments. Heating storage tanks and installing heat tracing on transfer lines is an option for addressing this challenge, but if that equipment is not already in place, then this option can carry substantial capital costs. An alternate option is to treat the stream with a PPD. Very simply, these are chemical additives that lower the pour point of a sample by frustrating the crystal packing behavior of the hydrocarbons in the sample so that the liquid will remain fluid down to lower temperatures.

Two examples of this are shown in Figure 5. In these cases, the untreated pyrolysis oils exhibited a base (untreated) pour point of 6-10 °C. While liquid at room temperature, exposure to even moderately cold temperatures would be likely to cause the streams to solidify. Treatment with a PPD had a significant impact on the cold flow properties. At a dosage of 600-1000 ppm, the pour point was able to be depressed to -21 °C or below. There is usually a limit to how much the pour point of a sample can be depressed, but we anticipate that in both cases a customer would experience fewer challenges associated with handling these samples in cold weather conditions, without the need to invest capital in heating equipment. These favorable laboratory results have been extended into successful field trials of the chemistry by various pyrolysis oil producers.

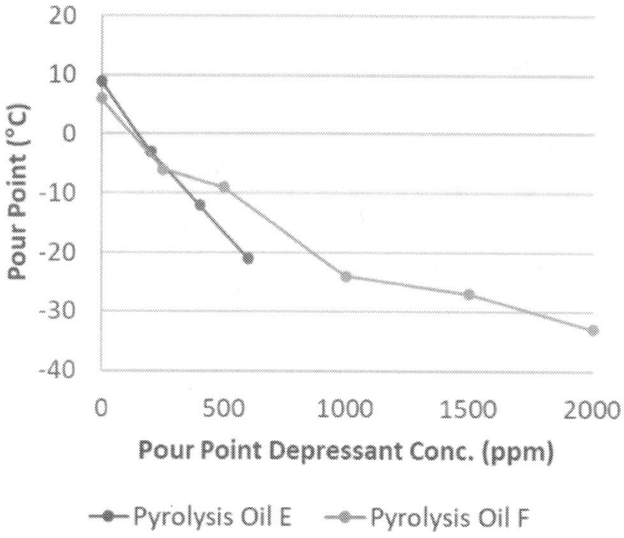

Figure 5. Impact of a pour point depressant (PPD) on the observed pour point of representative pyrolysis oil samples.

4 Conclusions and Outlook

Chemical recycling of plastic waste is a relatively old technology, but the high degree of interest in circular economy solutions from governmental groups, private companies, and the general public is enabling a period of high interest and potential growth for this industry. Nalco Water is working to partner with both PTF producers and users, and to that end we have introduced a family of products designed to address a variety of fundamental challenges associated with pyrolysis oils. We anticipate these offerings will provide timely and constructive solutions for both producers and users who are seeking to expand PTF technologies and processes. Nalco Water remains committed to this industry, as part of a broader strategy in circular economy solutions, and further offerings will be introduced as the field continues to progress.

5 References

[1] *EU restrictions on certain single-use plastics* (2021) https://ec.europa.eu/environment/topics/plastics/single-use-plastics/eu-restrictions-certain-single-use-plastics_en . Accessed 10 March 2022.

[2] *Single-use plastic: China to ban bags and other items* (2020) https://www.bbc.com/news/world-asia-china-51171491 . Accessed 10 March 2022.

[3] *Government of Canada moving forward with banning harmful single-use plastics* (2021) https://www.canada.ca/en/environment-climate-change/news/2021/12/government-of-canada-moving-forward-with-banning-harmful-single-use-plastics0.html . Accessed 10 March 2022.

[4] *Governor Newsom Signs Legislation to Tackle Plastic Pollution, Promote a More Sustainable & Renewable Economy and Protect Californians from Toxic Chemicals* (2021) Governor Newsom Signs Legislation to Tackle Plastic Pollution, Promote a More Sustainable & Renewable Economy and Protect Californians from Toxic Chemicals | California Governor . Accessed 10 March 2022.

[5] *A Circular Solution to Plastic Waste* (2019) https://image-src.bcg.com/Images/BCG-A-Circular-Solution-to-Plastic-Waste-July-2019_tcm9-223960.pdf . Accessed 10 March 2022.

[6] Smith M.D., Pickett T., Bernard A., Lally J. Plastics and the Circular Economy: Waste to Cracking Feedstocks. *AIChE Spring National Meeting, Ethylene Producers Conference*, 2020, Paper 159c.

[7] *Chemical Recycling 101* (2021) https://www.bpf.co.uk/plastipedia/chemical-recycling-101.aspx . Accessed 10 March 2022.

[8] *Alliance in Action: Progress Report 2021* (2021) https://endplasticwaste.org/en/our-stories/progress-report-2021 . Accessed 10 March 2022.

[9] *Circular Plastics Service* (2021) www.ihsmarkit.com/circularplastic . Accessed 10 March 2022.

[10] *ExxonMobil to build its first large-scale plastic waste advanced recycling facility* (2021) https://corporate.exxonmobil.com/News/Newsroom/News-releases/2021/1011_ExxonMobil-to-build-its-first-large-scale-plastic-waste-advanced-recycling-facility . Accessed 10 March 2022.

[11] Chevron Phillips Chemical completes first commercial sales of Marlex Anew circular polyethylene (2022) http://baytownedf.org/news-media/article/chevron-phillips-chemical-completes-first-commercial-sales-of-marlex-anew-c . Accessed 10 March 2022.

[12] Buekens A. Introduction to Feedstock Recycling of Plastics. *Feedstock Recycling and Pyrolysis of Waste Plastics*; Scheirs J., Kaminsky W. (Editors), John Wiley & Sons Ltd., Chichester 2006; pp. 3-41.

[13] Butler E., Devlin G., McDonnell K. Waste Polyolefins to Liquid Fuels via Pyrolysis: Review of Commercial State-of-the-Art and Recent Laboratory Researcher. *Waste Biomass Valor.* 2011, **2**: 227-255.

[14] Zhao D., Wang X., Miller J.B., Huber G.W. The Chemistry of Polyethylene Pyrolysis: A Process to Produce Fuel and Chemicals. *ChemSusChem* 2020, **13**: 1764-1774.

639302 Hydro-PRT - the Differentiated Plastic Waste Recycling Technology from KBR

Abstract:

Increasing global production and consumption of plastics is leading to accumulation of waste plastic in landfills and the environment, leading to societal, ecological and economic imbalances.

Advanced recycling technologies such as KBR's Hydro-PRT^{SM} process end-of-life and contaminated plastic waste, that can be further refined to produce virgin polymers and chemicals. KBR's Hydro-PRT is a differentiated, stable and scalable technology, based on a unique patented process using supercritical water to convert plastic waste into commercial products that can be used to produce new polymers delivering a truly circular economy.

The uniqueness of Hydro-PRT process offers following advantages:

High-product yields: no char or unwanted by-products, combined with the ability to control reaction conditions contribute to high liquid products yields.

Feedstock versatility: ability to process mixed composite polymers, flexible and rigid plastics and insensitivity to organic contaminants such as paper, card and biomass means the Hydro-PRT process can process a wider range of waste plastic feedstock than other competing technologies.

Scalability: direct heat transfer between the plastic and the heating medium (supercritical water) leads to an efficient heat transfer mechanism. Scale up is a simple process, not affected by heat transfer considerations.

Product stability: the water surrounding the plastic in the hydrothermal reactors acts as a suppressant to unwanted free radical reactions giving a higher yield of a far more consistent, stable product.

Product quality: the distillation of products allows for ready separation of fine solids and particulates into the heavy wax residue and water from the naphtha fraction. This provides high quality product oils (naphtha, distillate gas oil and heavy gas oil) with low solids and water content. Low chloride content, with most chlorides partitioning to the water phase rather than product oils.

Thermal efficiency: the process operates at lower temperatures requiring lower thermal energy input and the product gas can be reused in the process boiler to generate the supercritical steam providing a highly energy efficient process.

Operational flexibility: Hydro-PRT has a high degree of process control allowing products with different distillation ranges to be produced. Operators can bias products by operating parameters and blend of plastics in the feedstock.

Modular Offering: a modular offering of the technology brings cost and schedule benefits to any project.

Well-studied and demonstrated technology: the technology has been successfully demonstrated on pilot scale since 2014. A 20 thousand ton per year commercial scale plant utilizing Hydro-PRT is under design and construction by ReNew ELP in Wilton, UK. Using end-of-life plastic as feedstock, this plant will produce liquid hydrocarbon cracker feedstock alongside intermediate products. This plant will eventually process over 80 thousand tons of plastic waste feedstock per year and is scheduled to start-up in 2022.

Hydro-PRT provides clear environmental benefits:

- independent studies have demonstrated that advanced recycling technologies such as Hydro-PRT can reduce CO_2 emissions by 1.5 tons for every ton of plastic waste processed when compared to incineration.

- utilizing Hydro-PRT process to recycle plastics into new polymers leads to reduction in the use of fossil-sourced feedstocks.

- when recycling plastics waste via the Hydro-PRT process, minimal waste is produced -- impurities (colorants, additives, fillers etc.) in the plastic feedstock fall out into the heavier hydrocarbon feedstocks, which can be used in construction.

Benefits for business include reaching sustainability goals, developing sustainable infrastructure and reducing emissions and environmental footprint.

Hydro-PRT products resemble natural gas and oil with key applications ranging from making sustainable fuels to being refined into truly circular monomers, like ethylene.

The products, separated into fractions of choice based on their boiling points, are suitable for further processing in different types of crackers.

639523 Biodegradable Chitosan Composite Films from Banana and Potato Food Waste

Abstract:

The accumulation of plastic waste in the environment continues to be a growing concern worldwide. Additionally, a demand for more natural, bio-based products has opened a new market for biodegradable/compostable films for packaging across applications such as food and beverage, healthcare, personal care, and consumer goods. Developing suitable packaging materials from bioderived and biodegradable materials will reduce dependency on fossil fuels, lower harmful emissions from fossil fuel production, and contribute towards the development of a more sustainable and cost effective circular economy.

This study aims to improve existing formulations and food waste extraction methods to synthesize and produce alternatives to fossil fuel-based food packaging. Chitosan is a polysaccharide obtained from crab and shrimp shells extractions, which is also a known film-forming material. We synthesized biodegradable blend films using banana and potato food waste extractions and chitosan as starting materials, tested the mechanical, thermal, and water vapor permeation properties of the films. We also carried out the biodegradation testing in freshwater, soil, and compost environments to determine the fate of new biofilms in the receiving environments. These blend films reached 50-55% biodegradation in freshwater (day 91), 100% biodegradation in soil (day 45), and 50% biodegradation in compost (day 45). Therefore, the films demonstrate both high biodegradability in multiple environments as well as the physicochemical properties needed for potential use in single-use packaging applications.

639672 Techno-Economic Analysis of Waste Plastic Gasification to Methanol Process

Abstract:

Gasification represents a potential process to convert waste plastics to produce value-added products. One potential advantage of gasification is that it is mostly feedstock agnostic – namely, the plastic chemistry is not as important as it is for other recycling and upcycling technologies. This work reports a detailed techno-economic analysis (TEA) with Aspen Plus and supply chain analysis via Materials Flows through Industry (MFI) of the waste plastic gasification process to produce methanol. Steam gasification was modeled to convert the solid feedstock into syngas, which is cleaned and modeled with a Cu-based catalyst to produce methanol. Sensitivity analysis was performed to investigate the effect of various process parameters on the methanol Minimum Selling Price (MSP).

Our analysis indicates that methanol produced by waste plastic gasification exhibits an MSP substantially higher than fossil-fuel-derived methanol. The main driver of the methanol MSP is the feedstock cost, which is assumed to be $0.66/kg in our analysis. Sensitivity analysis indicates that cost-parity with fossil-fuel methanol can be achieved if the waste plastic feedstock is available at very low prices; materials recovery facilities (MRF) may have some waste plastic streams available at this price. Overall, this work identifies the important parameters that have the potential to reduce the MSP of methanol produced by waste plastic gasification.

640042 Perspectives of the Circular Economy for the Plastics Industry

Abstract:

Perspectives related to the development of a circular economy for plastics are presented. Plastics play an essential role in our lives, especially in the area of packaging. Plastic waste entering into our environment is unacceptable. Plastics must be properly managed and designed so as to make it readily recoverable from waste streams for recycling and reuse. Remember, a waste is a waste only when it now longer serves a useful purpose. If we can recover, recycle, and reuse plastics, they no longer are a waste.

A circular economy restores and regenerates by design. As a result, materials constantly flow around in a "closed loop" system, rather than having a single use and then being disposed of. Therefore, the value of plastics are simultaneously valued in the economy without leakage into the natural environment.

Most industrial process are conducted in a linear economy fashion. Processing begins with raw materials and a design; afterwards, the materials move to production and manufacturing. After production, the product moves to distribution and from distribution to consumption. As consumers, we can use this product for many years. After using it until we no longer have a purpose for it, we then dispose of it. After disposal, waste is collected and what is salvageable for recycling is moved to the next step, and what is not reusable will leave the cycle and undergo disposal into landfills, etc.

In the circular economy, the economy is circular using a "repair, reuse, and recycle" philosophy which is more friendly to the environment. This technique reduces wastes to a minimum and keeps whatever materials that are usable. As a result, using recycled materials keeps further value in these products, and minimizes waste and raw material consumption.

Key ideas associated with the plastics circular economy include: reinventing packaging and sorting practices; having better informed consumers; development of policies to help drive the change to a circular economy; and collaboration across sectors developing collaboration with industries, government, associations, etc.

Benefits of a circular economy are many. The circular economy uses and reuses raw materials efficiently and in many cases, fully. This helps ensure that there is less waste with more sustainable approaches and less negative environmental impacts. This also reduces the carbon footprint associated with the use of plastics.

640048 Techno-Economics Analysis and Life Cycle Assessment for Pyrolysis of Mixed Waste Plastics

Abstract:

Catalytic fast pyrolysis (CFP) is a promising approach to convert mixed plastic waste stream into valuable chemicals. To guide research and development further towards commercialization of plastics pyrolysis, economics and environmental impacts must be considered. In this work, we present a conceptual process design with techno-economic analysis (TEA) and supply chain analysis using Aspen Plus and Materials Flows through Industry (MFI), wherein mixed plastics are thermally depolymerized, and the resulting vapors upgraded catalytically to produce a suite of aromatics, olefins, and pyrolysis – naphtha (py-naphtha).

The base case model predicts a minimum selling price (MSP) for BTX aromatics at a zero net present value (NPV) for an nth plant with a 30-year plant life. The credits from the sale of multiple co-products significantly lower the overall costs. Feedstock cost, solvent cost, and discount rate are identified as key economic drivers through sensitivity analysis. A multivariate analysis was conducted to reflect the impact of fluctuations of WTI crude oil price on the different product and co-products obtained in this work. As catalyst research progresses, two scenarios based on high olefins and aromatics production were proposed, both of which show a potential reduction in MSP vis-à-vis the base case. Overall, this study suggests that research should prioritize ways to increase yields of industrially relevant products, minimize solvent loading, utilize low-value polymers such as films and flexibles that are currently discarded, and facilitate integration with petrochemical refineries.

PYROLYSIS: FILLING THE GAP IN RECYCLING WASTE PLASTIC

Tommy John, Tommy John Engineering, Bandera, TX

Abstract

Recycling, or material recovery, is the key to sustainability. Depositing material in a landfill or combusting in a Waste to Energy (WTE) plant, as is a common practice for municipal solid waste (MSW), should be replaced with material recovery. Sustainable material recovery from waste streams requires deep sorting <u>and</u> reliable, efficient processes for conversion of sorted streams into high value products. The pyrolysis process extends the range of plastic that can be effectively recycled from waste streams. It is effective with streams that have more diverse polymer compositions and contamination than can be tolerated with other plastic recycling technologies.

Producing synthetic oil and gas that can be used for petrochemical feedstocks and fuel products by pyrolysis is not a new process but it is currently receiving a lot of attention. A continuous plastic pyrolysis process is relatively straightforward when compared to typical hydrocarbon processes.

No single process will optimally recover mixed plastic waste streams, but multiple processes can be combined with rapidly advancing sorting technology to create a robust circular economy that improves sustainability and minimizes waste in the environment. Pyrolysis of plastic waste will be an important part of that endeavor.

Background

Sustainable waste management requires robust Material Recovery Facilities (MRF) to replace landfills and Waste to Energy (WTE) plants. Landfills have historically been the lowest cost option for waste disposal. WTE is more expensive, but often used in regions with a high population density and limited land. Both practices are difficult to permit and face considerable public opposition. Processing the full MSW stream in a MRF will maximize recycling rates and improve public cooperation with reducing plastic waste in the environment. Gaining acceptance for anti-litter campaigns is easier than for recycling, which is more complicated.

Landfills require a lot of land, and are often in a less accessible location requiring transportation of waste for a considerable distance. Although they can be constructed and operated to minimize environmental impact, the capture of landfill gas is limited to 70 to 80% by the need to limit air ingress to prevent landfill fires, resulting in unavoidable but substantial emissions of carbon dioxide, methane, and VOC. The landfill gas can be converted to natural gas or electricity, but the rate is low and carbon dioxide emissions are not abated.

WTE plants generate electricity using a steam cycle, but are limited by corrosion and fouling constraints to lower steam temperature and pressure, which, along with parasitic losses from emission control equipment, limits the thermal efficiency to 20 to 25%. This is quite low compared to current combined cycle units that have an efficiency approaching 60%. Although WTE produces a high value product, the recovery efficiency is low, and capital investment is high.

Digital technology enables improved sensors and mechanisms including robotics thus continuing to rapidly improve the sorting technology used in MRF. High level sorting can direct components of the waste stream for recovery as high value products, so it is paramount to have processes that effectively recover high value products from the sorted materials. As with all energy and disposal facilities, a MRF based waste processing facility will require a large initial capital investment. In order to pay off the bonds or other financing of the investment, a high value for the recycled material must be collected in addition to a tipping fee. The value of the recycled material also aligns with environmental quality because, if a product has a high value, it is a scarce resource and/or it requires a lot of energy to produce.

Waste plastics represent a unique challenge because of the diverse composition, but an opportunity because of the high value of the sorted components. No single process will optimally recover mixed waste streams, but multiple processes can be combined with rapidly advancing sorting technology to create a robust circular economy that improves sustainability and minimizes waste in the environment.

Pyrolysis Process

Pyrolysis of plastic waste will be an important part of that endeavor. This is not a new process; there has been a substantial amount of research, pilot plants, and even a few commercial plants implemented over several decades. For example, three commercial plants operated successfully in Japan in the 1990's, but lacked scale to be economic during a period of low oil prices and were closed down.

Pyrolysis processes extend the range of plastic that can be effectively recycled from waste streams. It is currently receiving a lot of attention because of the need to reduce plastic waste in the environment and improve the sustainability of the manufacture of very useful materials. It performs best for hydrocarbon based plastics, but it is effective with streams that have more diverse polymer compositions and contamination than can be tolerated with legacy plastic recycling techniques such as mechanical recycling, by which the material is ground to a powder and then extruded into pellets. The main product is an oil stream with a composition that ranges from heavy oil to light naphtha that can be used as a refinery feedstock that is equivalent to petroleum fractions. It can also be used as liquid feed to an olefins plant. An important secondary product is a gas stream that is rich in light olefins. It can be injected into an olefin plant for complete recovery, if the location has pipeline access to such a facility. Other-wise the gas can be compressed and the C3+ components condensed for storage and shipment in pressure tanks.

Process Description

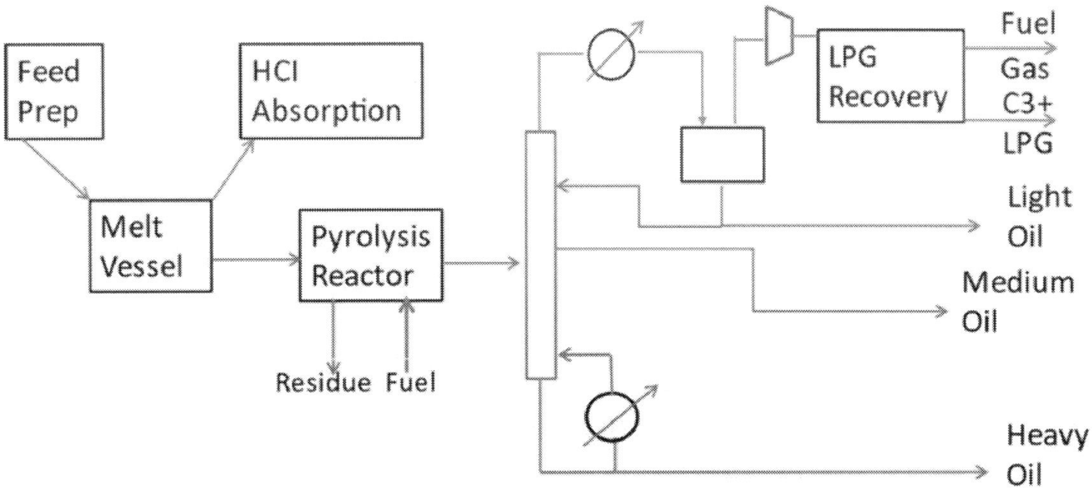

Polymer Pyrolysis Process

The solid feedstock is processed to a uniform size and is transferred into vessels where the material is melted and heated to 250-300 C with thermal fluid. At that temperature, the chlorine-carbon bonds break down in any chlorinated plastic, such as PVC, that has not been removed from the feed, and the chlorine is released as gaseous hydrochloric acid. The hydrochloric acid is absorbed into water and is neutralized and disposed of as a small waste water stream. The melted plastic is pumped to a reactor operating at 400 to 500 C and is cracked in a single step whereby all but about 5% by mass of the material is transformed into hydrocarbon vapors. Each reactor is heated by gas burners. The residue recovered from the reactor is coke or a very heavy pitch that can be blended with asphalt. The vapor from the reactor is quenched and passes to a distillation column where it is partially condensed and separated into three liquid products, light, medium and heavy. The distillation column overhead is water cooled and the bottom reboiler is heated with thermal fluid. The uncondensed gas is transferred to an olefin plant or compressed to about 21 bar (300 psi) and cooled and partially condensed as a hydrocarbon stream containing primarily light olefins and alkanes that goes to pressure storage and shipment. The uncondensed gas, composed of hydrogen, methane, and C2 hydrocarbons, flows into a fuel header that supplies most of the energy required for heating the reactors. Natural gas is made up into the fuel header to supplement the fuel supply. Circulating heat transfer fluid heated by the exhaust from the reactor burners supplies heat to the balance of the process. Electrical power is used for all mechanical drives, the largest usages being feed preparation, cooling, and compression. The largest air emissions will be from combustion for the reactors and fugitive emissions from process equipment and piping. The typical product distribution will be about 5% reactor residue (coke or asphalt), 10 to 30% gas (of which about 40% is light olefins), and the balance light, medium and heavy oil.

Value of Products

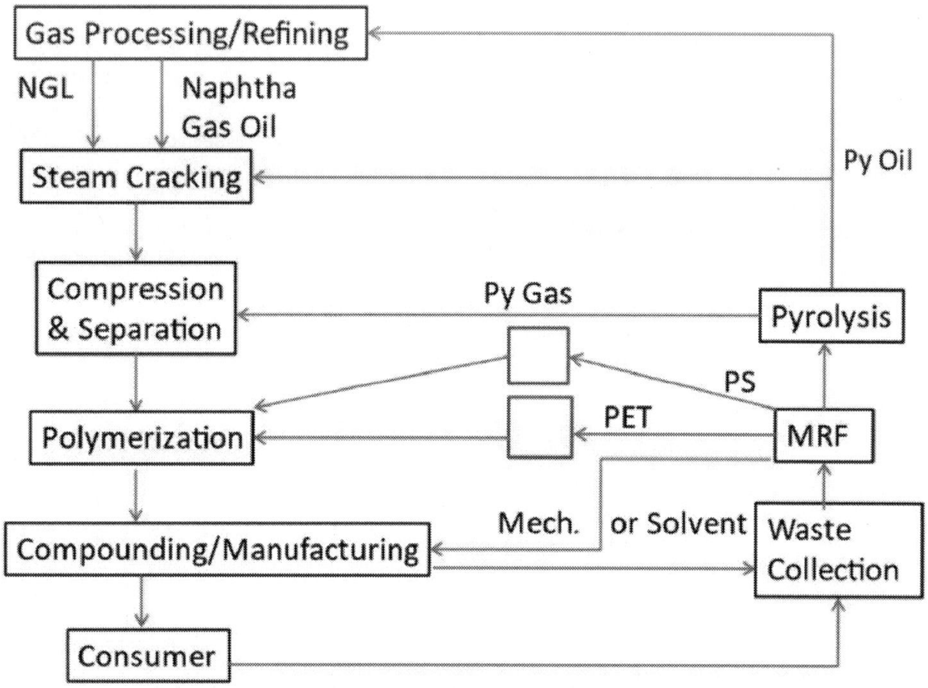

Hydrocarbon Polymer Cycle

The market value of the products produced by plastic pyrolysis will depend on the uses. The chart above is a simplified illustration of the hydrocarbon polymer supply chain for demonstrating the relative value and circularity of material from recycling processes. Pyrolysis complements rather than replaces Mechanical Recycling; the latter replaces polymer while the former is injected further upstream in the cycle. In the special case of polystyrene separated from the waste stream, it can be converted by pyrolysis into styrene monomer and used a feed to a polymerization process. Note that PET can similarly be recycled using chemolysis (developed by Eastman Chemical for example). The output of the Plastic Pyrolysis plant is py gas, py oil, and a limited amount of heavy residue (coke or asphalt, not to be confused with the heavy oil fraction).

In the special case of proximity to an olefins plant, the Py Gas can be fed to the compression section for full recovery of the stream. All of the constituents (alkanes, olefins, and hydrogen) are also in the cracked gas and will be recovered as product, feed, or fuel. Note that the pyrolysis plant does not have to be located within the olefins unit, but simply close enough for the gas to be transferred by a single pipe. The value of the py gas is the value of the individual constituents minus the cost of compression and processing. In the more common case of a remotely located Pyrolysis plant, the py gas can be compressed and cooled to partially condense as a C3+ hydrocarbon stream containing primarily light olefins and alkanes that can be shipped in pressured vessels using ground transportation. In this case, ethylene, which is about 10% by mass of the py gas stream, would have only fuel value. The cost of ground transportation would also reduce the value of the material shipped for recovery.

The use of Py Oil as a steam cracker liquid feed has been demonstrated. An alternative for any fraction that does not meet the specifications for steam cracker feed is use as input to a refinery. Compared to conventional crude oil, it is lighter, has more material in the range of transportation fuels, is low sulfur, and has lesser contaminants. Refineries have more options for processing and blending to meet fuel and feedstock specifications.

Inorganics in the waste plastic feed stream come from additives for imparting special qualities for specific applications. They are typically at very low concentrations, but are highly variable, and potentially create problems in applications for the product. Fortunately most inorganics in the feed will go to the small reactor residue stream, where is has limited impact. Some part of the contaminants will stay with the oil, so the specifications for steam cracker liquid feed need to be addressed with lab analyses. Specific items of concern are silicon, phosphorous, and metals.

Integration into Material Recovery Facilities

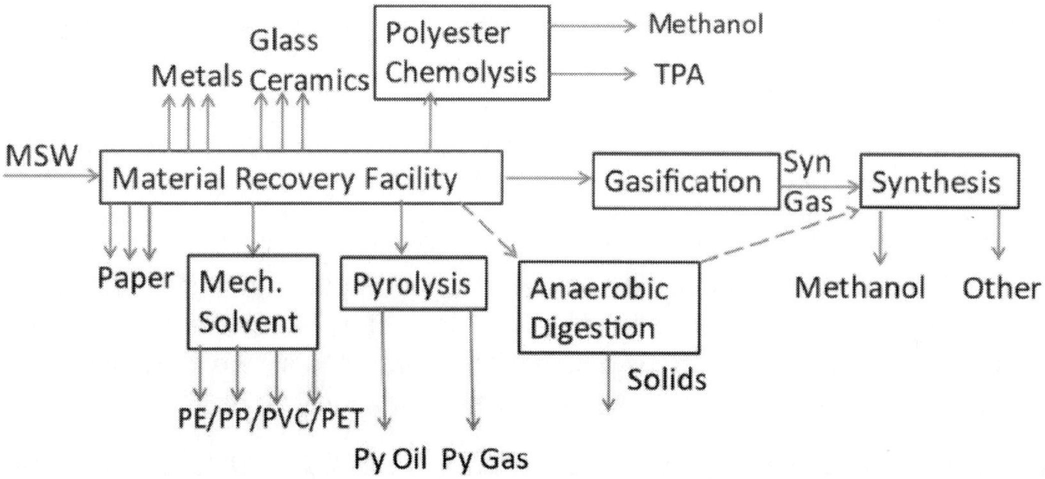

Material Recovery

State of the art sorting can capture materials from the whole waste stream for processing for the highest value, including multiple streams for metals, paper, and plastics. Dense solid materials such as glass, ceramics, and rocks have minimal commercial value, but are removed to improve downstream processing. The processing of plastic is challenging because of the range of products and the processes for recovery. Pyrolysis is particularly useful because it can convert many of the polymers remaining, after diversion into other processes, into Py oil, Py gas, and coke/asphalt.

The remaining materials, which may be more than half of the original waste stream, are mostly biomass derived or similar to biomass in composition: food waste, yard waste, wood, and uncaptured paper and plastic which can be fed to an oxygen blown, high pressure gasification unit for conversion to syngas composed primarily of carbon monoxide and hydrogen that can be used to produce organic oxygenates. Methanol is an example and has some government incentives, but it is a high-volume chemical produced with the economic advantage of economy of scale. Downstream oxygenated chemicals would have a higher inherent value because of additional processing and lower volume (less economy of scale for competitors). There are other options for conversion of the unsorted materials, including anaerobic digestions and composting, but would likely not create as much product value.

Conclusions

- Sustainable material recovery requires high level sorting and processes that efficiently recover high value materials.
- No single technology will optimize plastic recovery.
- Pyrolysis is a viable technology for increasing the amount of plastic waste that can be recycled.
- Material Recovery from the full waste stream is the best way to reduce waste plastic in the environment.

SPRING22
+18TH GCPS
A Joint AIChE and CCPS Meeting

Three Steps to Reinvent the Plastics Supply Chain

Ed Struzik, CPIM
Partner, Supply Chain & Product Stewardship
ERM
1920 Main Street | Suite 300 | Irvine, CA 92614
ed.struzik@erm.com

Prepared for Presentation at
American Institute of Chemical Engineers
2022 Spring Meeting and 18th Global Congress on Process Safety
San Antonio, TX
April 10 – April 14, 2022

Three Steps to Reinvent the Plastics Supply Chain

Ed Struzik, CPIM
Partner, Supply Chain & Product Stewardship
ERM
1920 Main Street | Suite 300 | Irvine, CA 92614
ed.struzik@erm.com

Keywords: plastics, supply chain, product stewardship, semiconductor

Abstract

The convergence of the global pandemic of 2020 and extreme weather events in 2021 created the "perfect storm" that exposed risks and lack of resiliency in large global supply chains. This paper addresses several alternatives to mitigate those risks in the plastics supply chain.

1 Introduction

The global pandemic exposed supply chain risks, particularly for industries reliant on Asian-based manufacturing and suppliers. Sources of semiconductor packaging, electronic final assembly, medical supplies, plastic components, and resistors were all disrupted. Many countries needed to find alternative supply chains to avoid halting operations. The semiconductor industry in North America became so concerned that the major fabricators, incentivized by the U.S. Government, committed to be more self-sufficient. Expanding sources of supply and near-shore material became the obvious strategy for minimizing risk and improving resiliency.

2 The supply chain storm

The plastics industry, with large amounts of automated, sustainable production anchored in the feedstock advantage along the U.S. Gulf Coast, seemed impervious to the disruptions of the pandemic. Companies produced materials to schedule until February 2021. Then a deep freeze hit the region, knocking almost 90 percent of production off line for over 30 days, costing the region at least $80 billion in lost revenue.[1] Suddenly the near-shore, automated supplier that was slowly recovering from pandemic disruptions was the problem, not the solution. Automotive, aerospace, heavy industrial, food and beverage, consumer goods, electronics, and pharmaceuticals were all now short of a key material. Pharmaceutical companies were in a particular bind due to the need for syringes to administer the COVID-19 vaccine.

No one can accurately predict force majeure like the pandemic and extreme weather. But it is possible to construct more resilient supply chains to prepare for catastrophic events. Companies can proactively reduce risks in three key ways:

1. Developing geographical diversification to reduce exposure to physical risk.
2. Creating sufficient buffer capacity in the way of critical material inventory.
3. Utilizing a combination of traditional suppliers and innovative companies that are creating circular supply chains.

And of course, each of these options must include measures to maintain or improve the Environmental, Social, and (Corporate) Governance (ESG) performance of the supply chain.

3 Diversify the supply base

What can industries do to offset the business risks associated with critical material shortages? The most obvious approach is to diversify the supplier base, as the electronics industry and others learned during the pandemic, by finding other suppliers for the key raw materials of feedstock and plastic.

Specifically, Asia Pacific and the Middle East have long had capacity and been suppliers. For example, electronics companies with plants in Malaysia, Singapore, and China could benefit from plastics production in that region. Similarly, European manufacturers could leverage capacity of Saudi Arabian plastics manufacturing.

However, as part of this reengagement process, the entire range of ESG and sustainability standards and audits must be addressed. New source of supply can only be brought into the mix for many of the consuming companies with robust auditing to ensure their current goals and programs are not set back.

For example, partnership with companies such as Ecovadis provides companies access to easy-to-use and actionable scorecards that equip procurement and supply chain teams with detailed insights into environmental, labor and human rights, ethics, and sustainable procurement themes across 200 purchasing categories and 160 countries. This allows supply chain teams the opportunity to assess any new sources to ensure they align with the goals set forth at their organizations.

4 Increase inventory to address shortages

Building up inventory would force plastics manufacturers to shore up their materials to a level suitable for use or conversion in a variety of industries. These stockpiles would require storage and conversion facilities, and the cost of carrying inventory would most likely be passed on to consumers. Developing these new inventory strategies would require in-depth coordination, analysis, and monitoring, likely adding overhead for the supplier and consumer. New storage and processing locations would come with increased transportation needs, and a direct increase in the carbon footprint and greenhouse gas emissions attributable to the supplier and the consumer of the material for ESG scoring purposes.

Increasing inventory levels improves resiliency and decreases risk, but at what cost? Some materials, like paint, mixers, resins, and coating bases, require special storage, labeling, and transportation. The plastic components for some industries like pharmaceuticals and food storage and packaging may have a longer shelf life, but the FDA has limitations on the shelf life for certain plastics used for food storage. These expiration dates will ultimately determine the movement and rotation of these stocks. The need to replace material outside of its approved shelf life would incur additional costs based on transportation distances and times, and increase greenhouse gas emissions.

5 Embrace the circular economy

Creating capacity, not in new production, but in recycling and circularity would significantly change the game for plastics manufacturers. The physical process, secondary or mechanical recycling—grinding, melting and reforming plastic material—is already widely used and accepted. A prominent example of this is plastic soda bottle manufacturing: the term "PET to PET" is common and recognized. However, this process often does not produce the purity standards required for certain polymers, polyvinyls (for paints), and electronics components. While many promising solutions in the mechanical recycling industry exist, there are some key challenges to really scale, including: improving primary and secondary sorting to better handle contamination, separating waste streams and materials that are more prone to mechanical processing, improving the overall efficiency of the process (e.g., automation) to reduce cost, and improving packaging design and materials (e.g., additives) to increase yield and minimize downcycling of the materials into less valuable products.

Adding capacity for chemical reprocessing, or tertiary recycling, involves the depolymerization, regeneration, and purification of monomers or oligomers. The regeneration process may require additional purification steps such as distillation, crystallization, and additional chemical reactions. This process shows great promise as a viable competitive source versus virgin production. However, while the technology and capacity are improving, effective scaling of the "waste" or return supply chain are struggling to scale the large collection and sorting infrastructure required to create a cost-effective supply.

Forward-looking companies with an eye on improving plastics circularity are already building this capacity. Firms such as BASF,[2] ExxonMobil,[3] Eastman Chemical,[4] and SABIC[5] are making the investments. These companies are providing their customers with an alternative supply. The waste plastic business is a $100 billion industry[6] and this change in mindset shows that plastic waste is a valuable alternative to virgin plastic. While these facilities do not insulate their customers from all force majeure (pandemics, deep freeze, etc.), they can be more geographically dispersed and do provide a buffer against the fluctuating price of virgin plastic. In addition, the costs of these facilities are relatively modest ($1 billion or less, which is insignificant when we consider the $80 billion in lost revenue from the deep freeze that occurred last year). Compared to the potential lost sales from interrupted supply, increased costs, and the reputational and ESG benefit to their customers of using a greater percentage of recycled material, this is an attractive solution for plastics consumers and recyclers.

6 Conclusion

The current recovery to supply chains is expected to take longer than previously thought and companies must make alternative business continuity plans to strengthen their supply chains in the future.[7] Many businesses are successfully re-evaluating their plastics supply chain by defining the risks and opportunities to increase capacity.

Many are also evaluating their programs and realizing the benefit of using recycled material as a key differentiator. This proactive strategy improves ESG scores and reputation, and prevents the operational shortages, price increases, and allocations that have become commonplace over the past two years. Undertaking a holistic assessment of the upstream supply chain to make it more resilient to supply disruption, and engaging with suppliers that can be agile, will be key in this evolution. All companies at some point will need to address these supply chain risks. The downside of not doing so may have an even more profound effect next time an event that impacts production occurs. Those that are able to think about opportunities gained by alternative measures now will reap the benefits later.

7 References

[1] The Federal Reserve Bank of Dallas. Cost of Texas' 2021 Deep Freeze Justifies Weatherization. https://www.dallasfed.org/research/economics/2021/0415.aspx. 2021. Accessed February 10, 2022.

[2] BASF. ChemCyclingTM: From plastic waste to virgin-grade products. https://www.basf.com/global/en/who-we-are/sustainability/we-drive-sustainable-solutions/circular-economy/mass-balance-approach/chemcycling.html. 2021. Accessed February 10, 2022.

[3] ExxonMobil. ExxonMobil tests advanced recycling of plastic waste at Baytown facilities. https://corporate.exxonmobil.com/News/Newsroom/News-releases/2021/0225_ExxonMobil-tests-advanced-recycling-of-plastic-waste-at-Baytown-facilities. February 25, 2021. Accessed February 10, 2022.

[4] Chemical & Engineering News (c&en). Eastman will build a $250 million plastics recycling plant. https://cen.acs.org/environment/recycling/Eastman-build-250-million-plastics/99/web/2021/02. February 1, 2021. Accessed February 10, 2022.

[5] SABIC. SABIC's circular solutions helping to address key sustainability challenges. https://www.sabic.com/en/newsandmedia/stories/our-world/sabics-circular-solutions-helping-to-address-key-sustainability-challenges. 2022. Accessed February 10, 2022.

[6] Material Handling & Logistics. Supply Chain Challenge: A $100 Billion Plastic Waste

https://www.mhlnews.com/global-supply-chain/article/21122896/supply-chain-challenge-a-100-billion-plastic-waste. 2020. Accessed February 10, 2022.

[7] Independent Commodity Intelligence Services (ICIS). US supply chain disruptions to take longer than anticipated to clear up – Fed. https://www.icis.com/explore/resources/news/2021/07/14/10663281/us-supply-chain-disruptions-to-take-longer-than-anticipated-to-clear-up-fed. 2021. Accessed February 10, 2022.

AUTHOR INDEX